Yellowstone Super Volcano…
Remnants Of A Celestial Impact

Richard L McGuire Jr.

CONTENTS

ACKNOWLEDGMENTS

To all the pioneers who have ever ask the question, Is there more? From Galileo, Newton, Christopher Columbus, Cortez, Lewis & Clark, Daniel Boone, Einstein, NASA, USGS, Geologists, Volcanologists. To all that have peered into the unknown and asked,… why these things? I hereby acknowledge that I would not have made this discovery if it was not for them. For it is their accomplishments and work that has made mine possible. All I have done is to take the work which they have produced and assembled the puzzle into the finished scene.

I have found that it was more difficult than I thought dealing with those in each field. I found them very protective of their view/field, reminiscent of the flat earth mentality that Christopher Columbus faced when he said the world was round. I asked one of the specialists why it was so hard to put out the data, and just have a look. His simple reply was... "funding". Something that competes with their funding is a threat to them. There is only so much funding available, even if yours is a completely different field. Only so much in the pool, and the competition is fierce. Which really dismayed me for awhile. All this work, and I was never going to get heard. All because the scientific community was more worried about funding, than shedding light on a scientific find, or truth that was not theirs. The other obstacle which was odd to me, was they spoke in absolutes. This is what we know, or think we know, so this is what it is. There is no absolutes when it comes to what was in the past, and what will come in the future, only theory. For space and earth are not static test tubes. The unexpected can and does happen, and will continue to do so. Do not be a flat earth proponent. Look, discover, and learn.

Even with that difficulty, it has been a wonderful adventure, seeking and then finding the answers I was looking for. I want to thank Dr. Robert B. Smith, research professor and professor emeritus of geophysics at the University of Utah and coordinating scientist for the Yellowstone Volcano Observatory, Dr. Peter H. Schultz Department of Geological Sciences, Brown University, Impact Cratering, and John D. West Graduate Student, Geophysics Arizona State University (ASU) School of Earth and Space Exploration (SESE), "Lithospheric Drip Beneath Great Basin", for taking the time on the phone, and email conversations.

I also want to acknowledge Matthew J. Fouch, Jeffrey B. Roth,and Linda T. Elkins- School of Earth and Space Exploration, Arizona State University, for their part in the discovery of the, "Great Basin Lithospheric Drip", NASA Visible Earth The Blue Marble, USGS, USGS - EnVision Energy Resource Program, and many more who's work has made this possible.

For it is their work, which makes up the pieces of the puzzle, and caused me to see that Yellowstone, is more than just a super-volcano.   RLM Jr.

Chapter One

Something Massive This Way Came

This book was written to show that Yellowstone Super-Volcano is more than just a ordinary volcano, super-volcano, or caldera. I am so confident that this is more than just a volcano, that with your indulgence, I am going to exercise a little bravado. And hereby claim the right of discovery to name this the, "Richard L. McGuire Jr. Yellowstone Impact Crater".

I intend to show that Yellowstone Super-Volcano is the remnants of a celestial impact, which created a crater in North America, comparable to the largest crater on the moon, being the South Pole-Aitken basin. This book also explains with a more persuasive logic, the cause and effects of the ongoing eruptions at Yellowstone. This book disputes the theory that the multiple eruptions, dating back in time from Yellowstone to the west into Nevada, were caused by the North America plate moving south westerly over a stationary hot spot below Yellowstone. But instead are eruptions that moved northeast along the asteroid or comets trajectory. With the northern most recent, "bubble eruption", being the Yellowstone Super-Volcano.

So lets begin...
Yellowstone Caldera is a remnant of a collision of the earth with a celestial body. Either an asteroid or comet of sufficient size, density, and velocity to penetrate the crust and enter the mantle of the earth. Yellowstone is near center of the crater, and northeast inline from the penetration point of the celestial body, a crater that has a diameter of approximately 2,190 Kilometers/1,361 miles. The celestial body having a trajectory from the southwest to northeast, penetrating deeper to the northeast. The object, as measured on the earths surface, having a sub-surface horizontal travel, from impact, to it's current 'bubble eruption' point, of approximately 603.5 kilometers/375 miles. Note, this is a horizontal distance from the point of impact to Yellowstone itself. This does not refer to the downward travel distance, which is dependent on the descent angle of the object, and can be determined with triangulation, once we have the correct angle. But from other sources later in the book, the estimate is greater than 400 miles deep and possibly half way to the earths core, 1,800 miles deep. Having a crater rim, to the west that rises, to 4,421 meters/14,505 ft./2 .75 miles above sea level, but then drops, in less than 123km/76 miles to the east, to a depth of -86 meters/-282 feet below sea level. Having a inner crater that covers an area approximately 3,719,716 km²/ 1,436,190 miles², and having a secondary outer crater rim, giving the total area to be approximately 4,149,279.7 km²/ 1,602,045.8 miles². Also to show that the celestial body upon penetrating the mantle entrained with it large volumes of gas, just as a bullet does when shot into a lake or pond. The subsequent bubbles rising through the mantle creating the line of multiple eruptions from Nevada to Yellowstone over the last 16 million years, showing that the eruptions are in-line with the trajectory of the object. Also to show a link between the end of the Cretaceous era, the KT boundary event and a possible link with the Chicxulub (pronounced chick-shoo-loob) crater. Suggesting a large asteroid or comet breakup similar to the Shoemaker-Levy Jupiter event of July 16, 1994, and compare with a potentially larger crater located on the north end of South America.

The basis for this discovery is purely observation of existing known facts, data and real world observations and is no different than any other scientific theory. The existing known facts and data will be compiled from NASA, USGS, and various scientific works and/or articles. For it is really the work of others that has made this discovery possible. The pieces of the puzzle exist, but it has never been assembled into the whole picture. This is like the analogy of a needle in a haystack, wherein the needle is so small, it is difficult to locate in the haystack. But just the opposite is true here, wherein, the needle is so large, you cannot even see the haystack, nor recognize the needle, as a needle.

Known facts. The earth and moon are riddled with craters of various sizes and ages. The largest moon crater being the South Pole-Aitken basin. Approximately 2,500 kilometers/1,391 miles in diameter and 13 kilometers/8 miles deep, with a northeastern rim elevation of approximately +8 kilometers/4.97 miles and a basin approximately -6 kilometers/-3.73 miles. *"Simulations at near vertical impacts show that this basin should have dug up vast amount of mantle material from depths as great as 200 kilometers/124.27 miles below the surface. However, observations thus far do not favor a mantle composition for this basin, and crustal thickness maps seem to indicate the presence of about 10 km of crustal materials beneath this basin's floor. This has suggested to some that the basin was not formed by a typical high velocity impact, but may instead have been formed by a low velocity projectile that hit at a low angle about 30° or less and hence did not dig very deeply into the moon. Putative evidence for this comes from the high elevations northeast of the rim of the South Pole Atkins Basin that might represent ejecta from such an oblique impact."*[1]

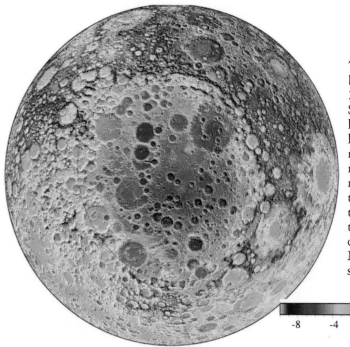

This LOLA image centers on the South Pole-Aitken (SPA) basin, the largest impact basin on the Moon (diameter = 2600 km), and one of the largest impact basins in the Solar System. The distance from its depths to the tops of the highest surrounding peaks is over 15 km, almost twice the height of Mount Everest on Earth. SPA is interesting for a number of reasons. To begin with, large impact events can remove surficial materials from local areas and bring material from beneath the impact craters to, or closer to, the surface. The larger the crater, the deeper the material that can be exposed. As SPA is the deepest impact basin on the Moon, more than 8 km (5 mi) deep, the deepest lunar crustal materials should be exposed here. In fact, the Moon's lower crust may be revealed in areas within SPA: something not found anywhere else on the Moon. [1]

-8	-4	0	4	8	Height/Depth Bar In Kilometers

Image credit: NASA/Goddard

To see a list of earth impact craters see wikipedia : List of impact craters on Earth

On July 16, 1994 the world witnessed Shoemaker-Levy comet breakup into 21 discernible fragments up to 2 km/1.24 miles in diameter. The comet took seven days from the first impact to the last, all within a very narrow band/latitude. The impact points made blasts two to three times the diameter of the earth.

Logic would indicate that if the Moon and Jupiter have been subjected to impacts of these dimensions, then why not the Earth? Where are these large craters? How would the remnants of such a collision be different on the earth than the moon? How would plate tectonics, volcanoes, weather erosion, and simple scale of size effect our ability to recognize a celestial impact after some several million years?

Image credit: NASA/Hubble Telescope

Chapter 2

Puzzle Piece #1 - The Crater

For several years I have been looking for a celestial impact crater, that I believed was responsible for the past eruptions leading up to the most recent Yellowstone eruption. I was looking in and around the Yellowstone Caldera. Tracing the path of the eruptions back in time into Nevada. All the while looking for a potential impact crater via satellite images. If I was going to find a impact crater related to Yellowstone, it was going to be big enough to be seen by satellite imaging. For years I was looking along the path of the eruptions. For I reasoned that the objects impact crater would be inline with the path, of past eruptions.

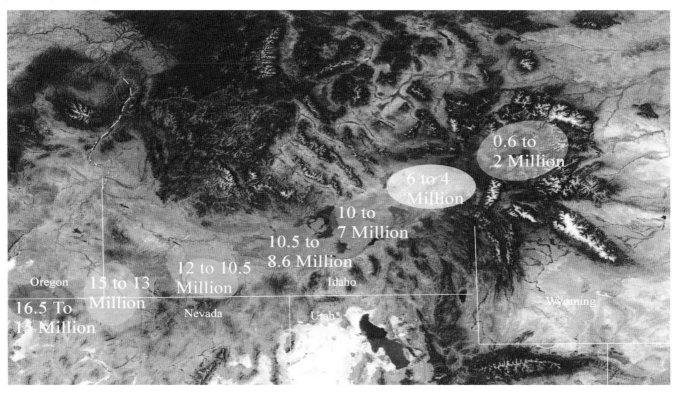

I knew that the impact site would have to be to the southwest of the oldest eruption. Ballistics would make that a reasonable assumption. Just as a bullet travels in a line, so would a high velocity, dense object penetrating the earth. So what was I looking for in the way of a impact crater? Lets look at the physical features of a impact crater, which is different than a volcanic eruption crater. A impact crater is going to have rim and possibly multiple rims, a blasted out basin, a central rebound or uplift, a direction/angle of travel, possible humping of the earth on the far end of the impact depending on the angle of attack. Volcanic craters, which result from explosion or internal collapse can look similar to impact craters but usually the terrain is built up from below, whereas impact craters typically have raised rims and floors that are lower in elevation than the surrounding terrain. Impact craters can be small, simple, bowl-shaped depressions to large, complex, multi-ringed impact basins with a rebound central uplift.

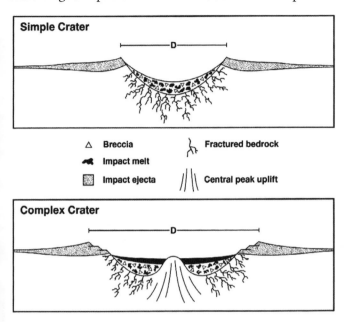

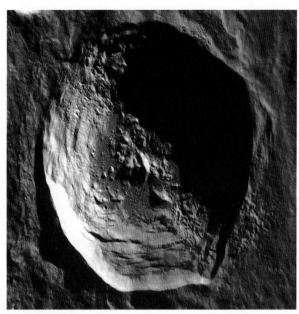

Image Credit NASA - Both the diagram above and the crater image on the right are showing impacts with a near vertical attack angle.

Image Right - Anaxagoras Complex Crater on the moon. Note the central mountain range rebound uplift. Also note the crater rim, mountain range in the lower left which is higher than the surrounding moon surface as seen by the shadow. The image has the sun shining from the upper right, and can play a optical illusion with your eyes. Upper right shadow is down inside the crater, central peaks are casting shadows towards the lower left.

The image below left is what one would expect to see from a oblique angle of attack from a high velocity impact. The earths rock would behave like clay, gelatine, or even liquid with a high velocity impact. Anyone who has ever shot a bullet into mud or clay will recognize a crater like this. The larger the ballistic, the larger the crater. The denser the object and higher the velocity, the greater the penetration.

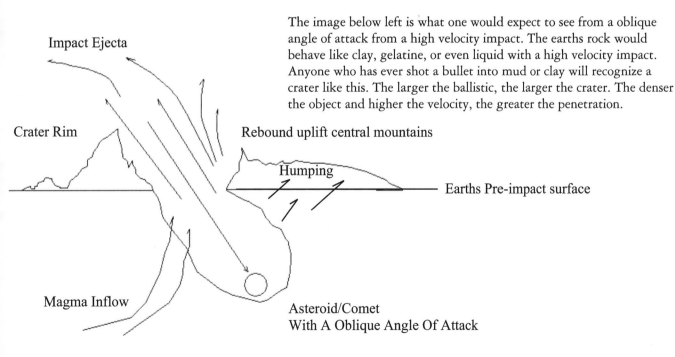

5

"When an object from space hits the Earth…
* There's a huge explosion.
* The impact makes a crater with a raised rim and sometimes a central peak.
* The hole is many times larger than the impacting object.
* There is a rapid release of a tremendous amount of kinetic energy as the object comes to a stop in about one hundredth of a second. (In the case of a hyper-velocity, very dense object - a plasma jet will penetrate ahead of the object and bore into the earths crust, just as a armor piercing round cuts through steel.)
* The impact releases extreme heat. Usually, the object itself is vaporized. Sometimes it melts completely and mixes with melted rocks at the site.
* If the impact occurs in water, a whole column of water is vaporized.
* The impact also produces a super-hot blastwave – a shockwave – that radiates rapidly outward from the impact point through the target rocks at velocities of a few kilometers per second.
* The shockwave is stronger than any material on Earth. It deforms rock in ways that are characteristic of an impact event. No other event on Earth deforms rock in these ways.
* Tiny glass droplets can form during the rapid cooling of molten rock that splashes into the atmosphere.
* Large impacts also crush, shatter, and/or fracture the target rocks extensively beneath and around the crater.
* Hot debris is ejected from the target area, and falls in the area surrounding the crater. Close to the crater, the ejecta typically form a thick, continuous layer. At larger distances, the ejecta may occur as discontinuous lumps of material.
* Large impact events can blow out a hole in the atmosphere above the impact site, permitting some impact materials to be dispersed globally by the impact fireball, which rises above the atmosphere. The resulting extensive dusk and smoke clouds can cause darkness lasting for a year.
* Special carbon molecules called Buckminsterfullerene or (Bucky-balls, after Buckminster Fuller) can travel to the earth in the impactor. They can hold special gases called "noble" gases that are indicators of extraterrestrial origin.
* Large impacts can trigger earthquakes and initiate volcanic eruptions.
* The heat ignites fires, and they may rage across a large region.
* Impact events can alter the chemical composition of the atmosphere. The extreme heat can generate large amounts of nitrogen oxides (NOx). NOx is easily transformed into nitric acid, resulting in acid rain". [2]

Now that you know what I was looking for in a crater, somewhere southwest of Yellowstone, and over into Nevada. The impact crater would have to be southwest of the oldest eruption in the chain of eruptions leading away from Yellowstone. That eruption was somewhere around 16.5 million years ago and is near McDermitt on the Nevada - Oregon state line. I happened to be looking southwest, trying to see any features that would indicate a crater. I was beginning to wonder what would be left of a crater that would have to pre-date the 16. 5 million year eruption. With tectonic plate movement, weathering, erosion, and maybe volcanic back flow that might have filled in the crater after impact, was there even going to be any evidence of a crater left to see?

I just happened to zoom way out on the satellite image, trying to re-orientate where I was on the image, and I gasped at what I saw. Right there before me was what I had been looking for. I was shocked at the massive impact crater that lay before my eyes. For I never expected to see a crater as enormous as this. Let alone, one that had not been discovered by the experts from NASA, Geologists, Oil exploration, etc., but I can also understand. The magnitude of what I was seeing, the impact crater being like my analogy at the beginning—— **Where the needle in the haystack was so large, that you could not even tell it was a needle… let alone see the haystack.**

When I zoomed out from looking in close in Nevada, here is what I astonished to find…………

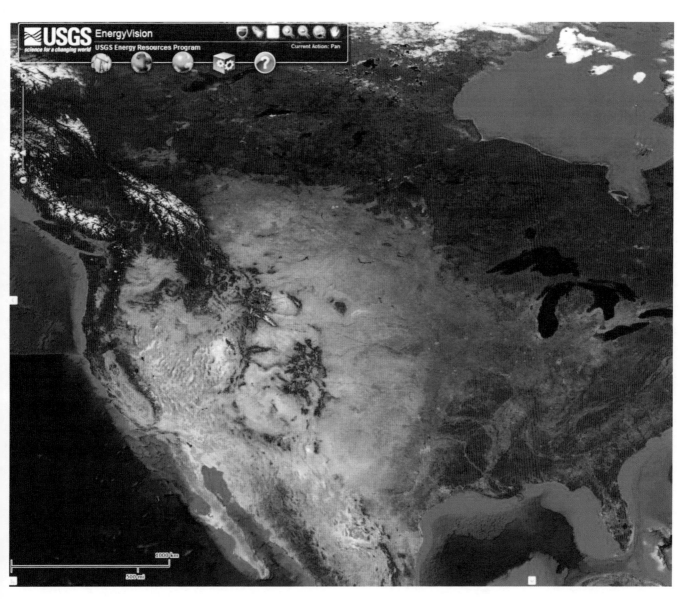

Image Credit USGS - EnVision

The above satellite image is what I beheld when I zoomed out. And only by zooming out was I going to find the massive impact crater that created Yellowstone. I was not prepared for a crater of this magnitude. Which is why I believe that no one else had discovered it. I have intentionally left this image as is, without notations so you can see for yourself the raw image for reference. You can also go to NASA's - The Blue Marble image collection and view the same image in greater detail. The photo's here are actual satellite photographs stitched together. These are not artist renderings, but actual images of the earth as seen from space. The impact crater is so massive that you would not be able to see it without this birds eye view. See the next image to see the crater notations.

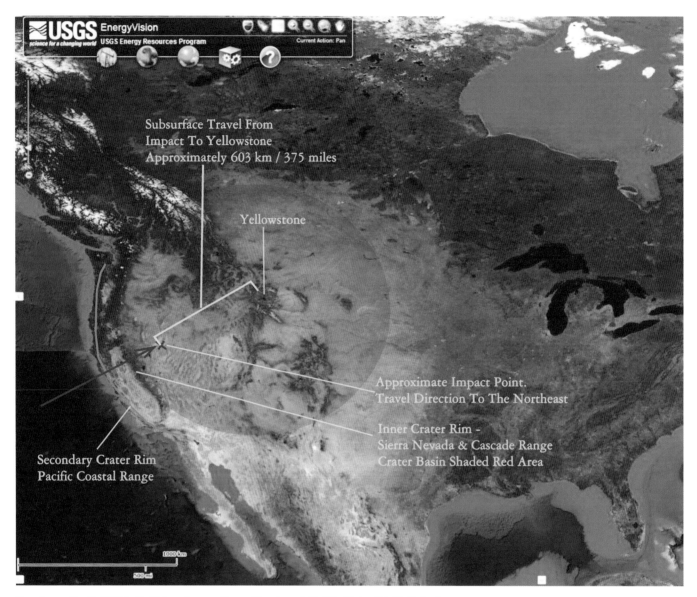

Base Image Credit USGS - EnVision - Impact Crater Notations Added By Richard L. McGuire Jr.

The above inner impact crater has a diameter of approximately 2,190 Kilometers/1,361 miles. The celestial body having a trajectory from the southwest to northeast, as indicated by the arrow, penetrating deeper to the northeast. With the object having, as measured on the earths surface, a sub-surface horizontal travel, from impact in Nevada, to it's current "bubble eruption" point at Yellowstone, of approximately 603.5 kilometers/375 miles. Note, this is a horizontal distance from the point of impact to Yellowstone itself. This does not refer to the downward travel distance, which is dependent on the descent angle of the object, which can be determined with triangulation, once we have the correct angle. But from other sources later in the book, the estimate is greater than 400 miles deep and possibly half way to the earths core, 1,800 miles deep. Having a crater rim to the west, Mt.Whitney, that rises to 4,421 meters/14,505 ft./2 .75 miles above sea level, but then drops, in less than 123 km/76 miles to the east, to a depth of -86 meters/-282 feet below sea level. The inner crater covers an area approximately 3,719,716 km2/ 1,436,190 miles2, with a secondary outer crater rim, giving the total area to be approximately 4,149,279.7 km2/ 1,602,045.8 miles2.

From the above satellite image you can see a near perfect circle for the inner Sierra mountains all the way up the Cascade mountain range in Washington and then on into Canada. The same also for the secondary Pacific Coastal range from the Mexican boarder to the Olympics in Washington. Nature and plate tectonics do not account for double ring, near perfect circles. The only logical answer is this is a impact crater of enormous proportions.

CHAPTER 3

Puzzle Piece #2 - The Geology

Now that I had found the impact crater, was there other evidence to support that this was a celestial impact? Does the geology support this to be a impact site? To the west is a double crater rim. The inner crater rim made up of the Sierra Mountain Range all the way up through Washington's Cascade Range and further to the west, the secondary outer crater rim, the Pacific coastal Range. The east side of California's Sierra Mountain Range has a dramatic rise from the basin to the top of Mt. Whitney.

Mt. Whitney rises to 4,421 meters/14,505 ft. above sea level, but then drops, in less than 123 km/76 miles to the east, to a depth of -86 meters/-282 feet below sea level. The Sierra Nevada Mountains is a tilted fault block nearly 400 miles long. It's east face is a higher rugged multiple scarp, contrasting with a gentle western slope about 2° that disappears under sentiments of the Great Valley. The image below, is looking west, and shows the near vertical face of the east side of the Sierra mountain range. ——————— Image below : NASA's Terra Spacecraft View of Mt. Whitney, the Highest Point in the Contiguous United States. In this 3-D perspective view looking west, created by the Advanced Spaceborne Thermal Emission and Reflection Radiometer (ASTER) instrument on NASA's Terra spacecraft, the Alabama Hills appear in the foreground. With its 14 spectral bands from the visible to the thermal infrared wavelength region and its high spatial resolution of 15 to 90 meters (about 50 to 300 feet), ASTER images Earth to map and monitor the changing surface of our planet. ASTER is one of five Earth-observing instruments launched Dec. 18, 1999, on Terra.

The shear vertical face of Mt. Whitney in the image at left, is the same you see in other impact craters. Going from 14,505 ft./2.75 miles high to -282 below sea level in such a short distance, 76 miles, is unprecedented anywhere else in the world. This image shows a great deal of material in the foreground that looks to have slumped down from above. How much higher would the rim have been if the material in the foreground had not been brought down to the crater basin by weathering or collapsing?

Image Credit - NASA's Terra Spacecraft

Serpentine occurs in central and northern California, in the coast ranges, the Klamath Mountains, and in the Sierra Nevada foothills. Serpentine is considered by geoscientists to be the metamorphosed remains of magnesium rich igneous rocks, most commonly the rock peridotite, from the earth's mantle. Peridotite is the dominant rock of the Earth's mantle above a depth of about 400 km /248.5 miles. Beneath the oceans, the crust varies little in thickness, generally extending only to about 5 km /3.1 miles. The thickness of the crust beneath continents is much more variable but averages about 30 km / 18.6 miles; under large mountain ranges, such as the Alps or the Sierra Nevada, however, the base of the crust can be as deep as 100 km / 62 miles.

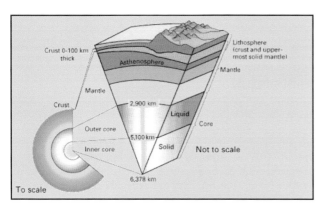

Image Credit - USGS

One must consider the forces needed to bring peridotite from the earths mantle to the surface over such a large area.

In the center of the crater, we have the Rocky Mountain Range, or rebound uplift from the impact. *" Geologists believe that the modern Gunnison River became established in its current course about 10 to 15 million years ago, just after the last eruptions in the San Juan's and West Elks. This coincides with the beginning of a period of rapid uplift of the Great Basin and Colorado Plateau provinces that lie between the Rockies and the Sierra Nevada Range in California . To date, geologists are at a loss to explain the forces behind the uplifting of such an immense region.... The Laramide orogeny was a period of mountain building in western North America, which started in the Late Cretaceous, 70 to 80 million years ago, and ended 35 to 55 million years ago. The major feature that was created by this orogeny was the Rocky Mountains.* "[3]

The fact that geologists are at a loss to explain the forces behind the uplift or rebound of such a large area in the middle of the North American Plate, gives further credence to a massive celestial impact as being the cause. And again when looking at the satellite image below, we have a central rebound, a crater rim, making a complex crater very evident.

Image Credit NASA - Annotations Added For Reference - Note: Sun is shining from upper right corner to lower left corner

Image Credit - USGS - EnVision - Annotations Richard L. McGuire Jr.

You could lay the Anaxagoras moon crater above, over the Yellowstone Impact crater left, and have an exact match of physical features. Plate tectonics create random events, not repeatable patterns and cannot explain the similarities between these two images. Impact craters, "*do create*", repeatable patterns and features.

One of the other attributes of a celestial impact here on earth would be geothermal activity from a large impact, due to the fact that the earth is geothermally active just below the crust. A large impact is going to blast away and fracture the crust, causing a greater amount of geothermal activity closer to the surface of the earth. That is precisely what we find. The crater is encircled by a diverse range of volcanic and geothermal activity. Not only encircling the crater, but throughout the crater itself there is a myriad of volcanic activity. Geothermal activity that is unprecedented anywhere else on the continental North America plate.

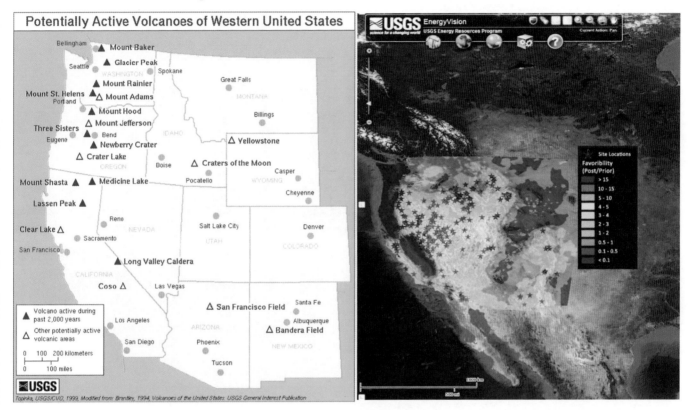

Image credit - USGS

Image credit- USGS - EnVision

As you can see in the two images above, geothermal activity is throughout the impact crater. In fact the image above right, USGS - EnVision energy resource map with the geothermal area overlay, does not even show any potential geothermal sites east of the crater. Also notice the wake effect of geothermal activity inline with the trajectory of the objects impact and travel straight towards Yellowstone. Evidence indicative of a major impact crater being blasted out, and after several million years, it is still geothermally active .

Another important geological feature from a physical impact with earth would be compressed seismic zones.

Image Credit - USGS with seismic zone overlay - Annotation added by R.L. McGuire Jr.

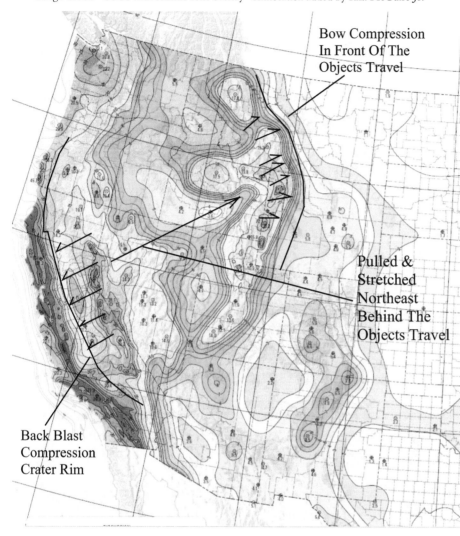

Bow Compression In Front Of The Objects Travel

Pulled & Stretched Northeast Behind The Objects Travel

Back Blast Compression Crater Rim

The image at left shows the impact crater seismic zones. The seismic lines show how the crust of the earth was stretched, pulled, and compressed by the impact. The tighter the lines are together the more compression of the earths crust.

You can see the path of the object very clearly, just as a boat would create a wake as it travels through a body of water. With the water being compressed at the bow, and pulled or stretched following the boat.

Note that there is no distortion or seismic compression east of the impact crater, (light purple areas).

With the North American Plate moving to the southwest as one giant slab, wouldn't you expect to see a consistent amount of compression to the south and east also? Based on conventional plate tectonics, if this was one slab, why the compression only in the west and around Yellowstone? Butting up against subduction zones off the pacific coast does not account logically for this distortion. I dispute the current theory that the bow wave around Yellowstone is because the North American Plate is sliding over a stationary hot spot. It may be a contributing factor, but it is not the main cause.

One can clearly see that a dense object traveling at great speed impacted the ground in Nevada, blasted a crater rim backwards, southwest out towards the pacific ocean. It pulled and stretched the earths crust northeast following the object, creating a bow wave of crust in front of the object as it moved towards Yellowstone.

"All over North America, the K-T boundary clay contains glass spherules, and just above the clay is a thinner layer that contains iridium along with fragments of shocked quartz. It is only a few millimeters thick, but in total it contains more than a cubic kilometer, (.24 cubic mile), of shocked quartz in North America alone. The zone of shocked quartz extends west onto the Pacific Ocean floor, but shocked quartz is rare in K-T boundary rocks elsewhere: some very tiny fragments occur in European sites. All this evidence implies that the K-T impact occurred on or near North America, with the iridium coming from the vaporized asteroid and the shocked quartz coming from the continental rocks it hit." [4]

The geological time line of the crater and the outside surrounding area is very significant in determining that this is a impact crater. Lets take a look at what one would find if there had not been a impact. Lets also have a look as if there were very minimal plate tectonics and weathering. Only enough weathering to lay down the layers throughout the ages. The image on the left is the USGS official geological time line as of 2006. If there was no plate tectonics, weathering, volcanoes, etc., we would have a perfect time line of the earths layers just like the image on the left. Layer on top of layer, a nice stack of pancakes. But the earth is not static. There is a myriad of forces acting upon the earths surface. Earthquakes heave the land. Volcanoes bring material from deep within the earth to the surface. Wind, rain, floods, waves, oceans, all come and go. Each changes the face of the earth.

So what would be different in the impact crater? For the most part lets consider that each time line was laid down in nice, neat, geological order as indicated to the left. In the crater, we would expect that nice neat layering to be blasted out, disheveled, layers upended, fault blocks upturned, and layers that lay deep down now exposed on the surface. In other words a very mixed, chaotic bag of geology, and the earth broken up. Note: The color bar at the left has the most recent at the top, oldest at the bottom.

Once we leave the blast zone, the further to the east we go, where the object bores into the earth, and humps the earth up without destroying the layers, the more order in the layers one would expect to find. With the newest layer laid down on top of old layers. Other than weathering, rivers cutting through, and volcanoes, it would be more uniform. The image below shows just that. All to the west, and in the crater area is a mixed bag of geological time, yet the further to the northeast, southeast, and east we go from the crater the more uniform the geology is. Even the geology shows there was a major upheaval from a major impact. To the west of the Rocky Mountains the geology is a mixed bag, while to the east a very uniform top layer. This image also seems to indicate massive outflows to the north into Alaska as well as to the south down into Mexico.

Image Credit - USGS - EnVision with geology overlay

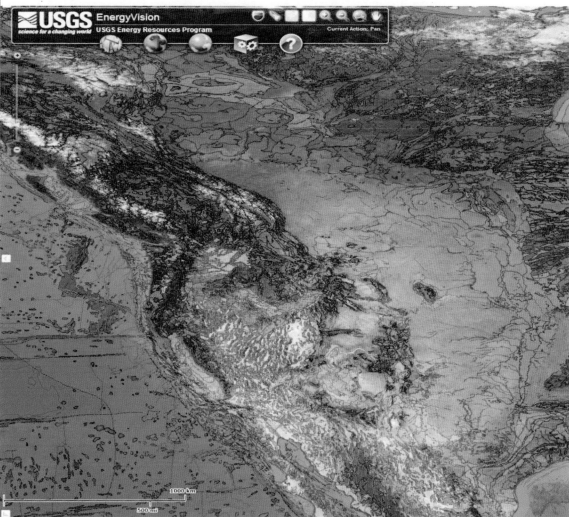

Puzzle Piece #3 - The Ballistics

The ballistics and trajectory of the impact are from real world observations. When I was young, I would often observe the moon through my telescope, and imagine the meteors crashing into the moon, and blasting out the craters. I would re-create those impacts in mud and on moss covered ponds by hurling rocks and boulders into the surface. I would also shoot my various rifles — BB gun, 22 rifle, 30/30 rifle, into the pond and surrounding mud. If I lightly tossed a small stone onto the moss, the stone would bounce and settle on top. But with a little exertion at a higher velocity the stone would penetrate the scum, and a trail of entrained bubbles would mark the path of the imagined meteor hit.

I would imagine the scum as the crust of the earth, as it buckled from the impact. Some of the bubbles would immediately blast to the surface. While other smaller bubbles would get trapped under the scum, and not break through the surface. If I tossed a large rock as vertical and high as I could, the rock would enter straight down like a cannon ball. Producing a wonderful vertical back blast, and all the entrained air would rise straight back up along the rocks entry point. But if I threw the rock hard at an oblique angle, I would get a back splash mirroring the entry angle and a nice trail of bubbles rising to the surface along the path my rock took. If thrown or shot into mud at an oblique angle, it would produce a half crater with the back splash, while ahead of the projectile as it buried itself, the ground would hump up over the path of the object. Just like the illustration at the bottom of page 4.

It did not matter the size or velocity, whether a rifle bullet, BB , or a rock, all would exhibit the same characteristics and would entrain air into the water behind the object. So based on real life observations, one can see a similar pattern for a celestial impact, but on a much grander scale for the Yellowstone impact. Consider for a moment a large hypervelocity, dense, even metallic object slamming into the earth at a oblique angle, just east of Mt. Whitney in Nevada. Immediately upon impact a huge blast erupts, but the object is not disintegrated by the impact. Because of it's density and velocity it punches through the earths crust and on into the mantle of the earth. At hyper-velocities you would have the earths surface behave like clay, gelatine, or even liquid. The projectile would also blast a super heated plasma jet ahead of it, just like a welders cutting torch, or like a armor piercing round melts through steel. That plasma jet in and of itself is part of the projectile vaporizing itself into a gas, as well as the earth being vaporized as the object penetrates deeper. Now just like the scum covered pond you have multiple, huge gas bubbles, under the earths crust, and entering deep into the earths mantle following the projectile. Just like the back splash on the pond, immediately after impact, a large volume of earths crust and molten lava blasts backwards out in the opposite direction of the entry channel. The crater boils, and rolls with molten lava for maybe millions of years before it settles down again. Which would cause unprecedented volcanism in the crater itself, and even perhaps on other sites around the globe, such as the Deccan Traps in India. More on that later…

But for now, what do we have that shows a hyper-velocity ballistic impact which was powerful enough to do as I described above took place? Well for starters lets look at my, 'bubble eruptions', scenario as being the cause for the multiple eruptions leading up to and including Yellowstone Super Volcano. Under current theory, the North American Plate is sliding over the Yellowstone stationary hot spot. And the past eruptions have moved southwest all the way into Nevada from the current hot spot in Yellowstone. From all the evidence I can see, that is not the case, but rather the eruptions traveled northeast following the projectiles path under the earths crust. And Yellowstone is the most current bubble of gas rising to the surface, hence the reason I call them, "Bubble Eruptions". The bubble eruptions further back in time were the ones closest to the earths surface and erupted first. The deeper the bubble, the longer it takes to reach the surface. If the gas bubble can out-gas fast enough through cracks and fissures in the ground there will be no eruptions. Just like a balloon inflating and deflating as it out-gases. But if the bubble rises faster than it can out-gas, then the balloon pops, and we have an eruption. As further evidence Yellowstone's magma chambers contain large volumes of gas. [5]

Below is the path of the eruptions going back in time towards Nevada. With the first eruption 16.5 to 15 million years ago. So based on the observations of ballistics, the first bubble to reach the surface would be the one closest to the surface, followed by the next deepest, and so on and so forth. As evidenced by the time line of the historical eruptions leading to Yellowstone, one can see that each succeeding eruption came at a later date and thus a deeper bubble reaching the surface.

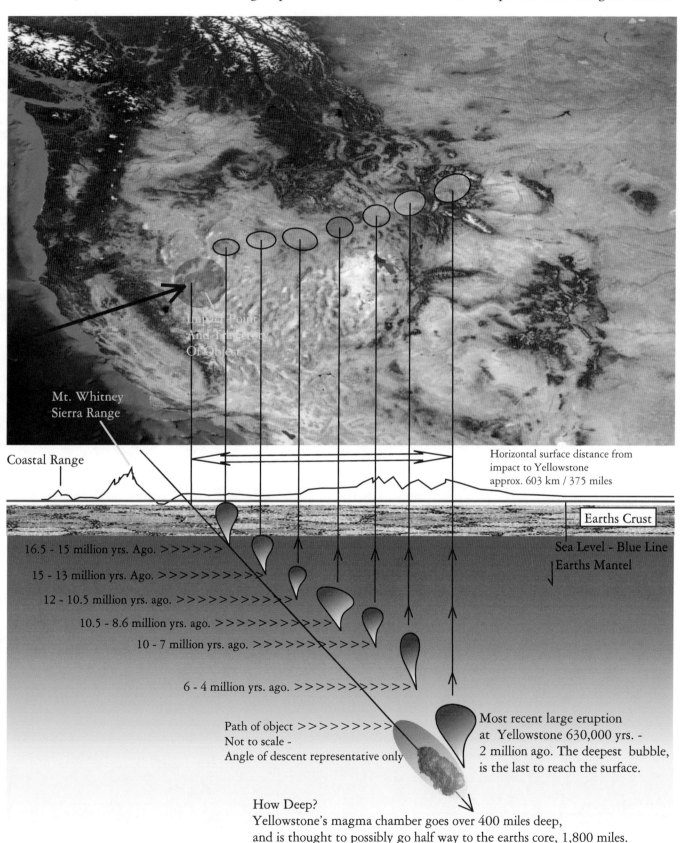

Mt. Whitney
Sierra Range

Coastal Range

Impact Point
And Trajectory
Of Object

Horizontal surface distance from
impact to Yellowstone
approx. 603 km / 375 miles

Earths Crust

16.5 - 15 million yrs. Ago. >>>>>>

15 - 13 million yrs. Ago. >>>>>>>>>>

12 - 10.5 million yrs. ago. >>>>>>>>>>>

10.5 - 8.6 million yrs. ago. >>>>>>>>>>>

10 - 7 million yrs. ago. >>>>>>>>>>>

6 - 4 million yrs. ago. >>>>>>>>>>>

Sea Level - Blue Line
Earths Mantel

Path of object >>>>>>>>>
Not to scale -
Angle of descent representative only

Most recent large eruption
at Yellowstone 630,000 yrs. -
2 million ago. The deepest bubble,
is the last to reach the surface.

How Deep?
Yellowstone's magma chamber goes over 400 miles deep,
and is thought to possibly go half way to the earths core, 1,800 miles.

Another indication of a ballistic impact would be back splash or magma eruptions, back flowing out of the crater for years after the impact. Possibly for millions of years, for a impact of this magnitude. Consider the earth as a large round spherical container. The surface of which is fractured, cracked, porous. In fact the only thing containing the inner liquid, "magma", is gravity, equilibrium and a thin shelled crust. Just like a M&M piece of candy. When you consider this, you also have to consider the laws of physics. One law of physics is you cannot compress a liquid, or a solid. You can only compress a gas. Any mechanical engineer will tell you when you compress a liquid in a sealed container you are going to have something give. Pistons break, o-rings and seals rupture. Think of any gas filled tank. You never fill that tank over 80% so as to leave enough headroom of gas, to allow for expansion from heat. If the tank is 100% full of liquid, any heat which causes the liquid to expand is going to rupture or even explode the container violently.

The earth is no different, only on a much grander scale. You squeeze it, it will squirt molten lava out from the weakest point on it's surface. In fact, I would be willing to wager that if one was to go and do a study of volcanoes since the Sumatra, Chile, and Japan earth quakes… one would discover a increase in volcanism and outflow. Which will continue until an amount equal to the mass that was displaced is reached. Consider the following quotation from NASA JPL News story dated 03/14/11……

"Japan Quake May Have Shortened Earth Days, Moved Axis 03.14.11

Using a United States Geological Survey estimate for how the fault responsible for the earthquake slipped, research scientist Richard Gross of NASA's Jet Propulsion Laboratory, Pasadena, Calif., applied a complex model to perform a preliminary theoretical calculation of how the Japan earthquake—the fifth largest since 1900—affected Earth's rotation….

The calculations also show the Japan quake should have shifted the position of Earth's figure axis (the axis about which Earth's mass is balanced) by about 17 centimeters (6.5 inches), towards 133 degrees east longitude. Earth's figure axis should not be confused with its north-south axis; they are offset by about 10 meters (about 33 feet). This shift in Earth's figure axis will cause Earth to wobble a bit differently as it rotates, but it will not cause a shift of Earth's axis in space—only external forces such as the gravitational attraction of the sun, moon and planets can do that.

Both calculations will likely change as data on the quake are further refined.
In comparison, following last year's magnitude 8.8 earthquake in Chile, Gross estimated the Chile quake should have shortened the length of day by about 1.26 microseconds and shifted Earth's figure axis by about 8 centimeters (3 inches). A similar calculation performed after the 2004 magnitude 9.1 Sumatran earthquake revealed it should have shortened the length of day by 6.8 microseconds and shifted Earth's figure axis by about 7 centimeters, or 2.76 inches. How an individual earthquake affects Earth's rotation depends on its size (magnitude), location and the details of how the fault slipped. Gross said that, in theory, anything that redistributes Earth's mass will change Earth's rotation."

In other words, the "Mass" of the fault blocks slipping inwards on the earths surface was enough to cause the earth to speed up it's rotation. Just like an ice skater speeds up when spinning, as her arms are pulled in close to the body, bringing the mass inwards. —— Now back to my wager,… If that amount of mass was moved inwards, due to the fault block slipping down, causing the earth to increase it's rotation speed… then according to the laws of physics, displacing that mass inwards, (compressing the mantel), is going to cause an equal amount of displacement elsewhere on the globe. IE: An increase in volcanoes and or earthquakes uplifting mass to equal the displacement downward from those earthquakes. That outward displacement will continue until an equilibrium is reached. So, with each massive earthquake, where a fault block slips 'downward', expect to see an equal displacement upwards. Simple law of physics.

Now back to our ballistics. You ever see what a bullet does to a gallon jug of water? Imagine the forces on the earth for a impact of this size. First you would have the back blast from the initial impact, followed by a massive back flow of lava and mantel flow. The crater floor would seethe and boil for millions of years. The earth from the compression of the impact would probably undergo world wide volcanism.

"Exactly at the K-T boundary, a new plume, "of volcanism"(sic), was burning its way through the crust close to the plate boundary between India and Africa. Enormous quantities of basalt flooded out over what is now the Deccan Plateau of western India to form huge lava beds called the Deccan Traps. A huge extension of that lava flow on the other side of the plate boundary now lies underwater in the Indian Ocean. The Deccan Traps cover 500,000 km2 now (about 200,000 square miles), but they may have covered four times as much before erosion removed them from some areas. They have a surviving volume of 1 million km3 (240,000 cubic miles) and are over 2 km thick in places. The entire volcanic volume that erupted, including the underwater lavas, was much larger than this. Furthermore, the Deccan eruptions began suddenly just before the K-T boundary. The peak eruptions may have lasted only about one million years (±50%), but that short time straddled the K-T boundary......

Thus there is strong evidence for short-lived but gigantic volcanic eruptions at the K-T boundary. Some people have tried to explain all the features of the K-T boundary rocks as the result of these eruptions. But the evidence for an extraterrestrial impact is so strong that it's a waste of time to try to explain away that evidence as volcanic effects. We should concentrate instead on the fact that the K-T boundary coincided with two very dramatic events. The Deccan Traps lie across the K-T boundary and were formed in what was obviously a major event in Earth history. The asteroid impact was exactly at the K-T boundary. Certainly something dramatic happened to life on Earth, because geologists have defined the K-T boundary and the end of the Mesozoic Era on the basis of a large extinction of creatures on land and in the sea. An asteroid impact, or a series of gigantic eruptions, or both, would have had major global effects on atmosphere and weather." [4]

That is the exact scenario that I believe this event caused. It was not just the impact that resulted in the K-T event, but the impact and volcanism of epic proportions combined. Why does the scientific community insist that it was one or the other that killed the dinosaurs? Why not both? The impact, then the resulting volcanoes combined to give a one-two knock out punch, ending the rule of the dinosaurs. The image below seems to show massive outflows of mantel material, known as magmatism, all the way north into Alaska and south into Mexico, (red arrows). With weathering over the millions of years leaving mountains, since the event, is there evidence that this is related to the physical anomaly seen? Are the mountain top's related to an outflow of material caused by the impact? That would take field geologists sampling, looking for a specific relation to the impact, and is well beyond my capabilities and resources. So I will leave that to others, more competent than I. But the physical features of what looks to be great flows north and south from the crater can easily be seen in the satellite image left…. As well as in the geological overlay image on the next page.

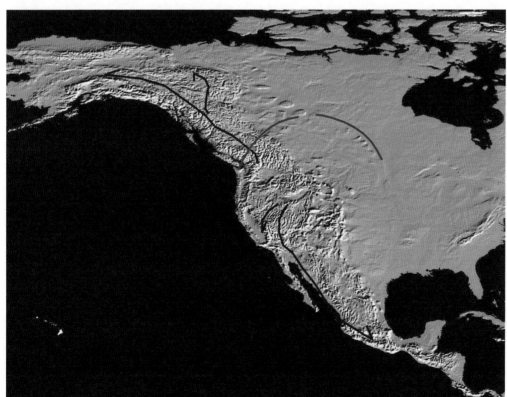

Note: In this shaded gray scale elevation image, you can see the northern, and eastern continuation of the crater rim under the humping as indicated just inside the blue line.

Image Credit- NOAA Elevation Shaded Relief Map

Even the geology below suggests that there was an out flow as indicated by the red arrows.

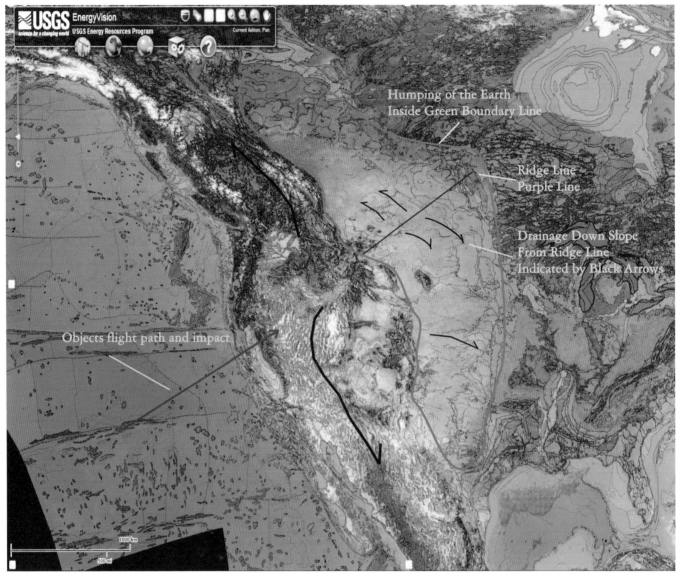

Image Credit - USGS EnVision Energy Resource Map With Geology Overlay - Annotations Added By R.L. McGuire Jr.

Another anomaly in the satellite image that indicates a celestial impact is the humping of the earth on the east side of the rocky mountains and extending well past the craters original rim. The humping extends north, east and south. All of the land slopes away from the central ridge, (purple line), which is in a direct line with the objects trajectory. Just as one would expect to find in a ballistic model, of a oblique angle impact crater. (See the last image on the bottom of page 4) One has to ask the question… Did the object travel northeast past Yellowstone? If so how far and how deep? If that is the case, then according to the downward and northeast trajectory, one could expect future eruptions northeast of Yellowstone. Only time will tell.

" Crater maximum depth depends on the impactor density, whereas crater(sic.), diameter depends on target density. As the impactor continues to penetrate, it transfers its momentum and energy to the target as it disrupts and deforms. In planar targets, this transfer depends on the response of the impactor to the initial shock created at first contact….. The rate of energy transfer depends on impact velocity and projectile/target impedance. Low velocity impacts or soft coupling result in pear-shaped craters with the apex pointing up-range. High velocity impacts or strong coupling result in a pear-shaped crater with the apex pointing down-range." [6]

The image below clearly shows a pear-shaped impact, with the apex pointed down-range. Which indicates a 'High Velocity' crater impact. Also see the image on p. 12, the geology also clearly shows a pear-shape with the apex down-range.

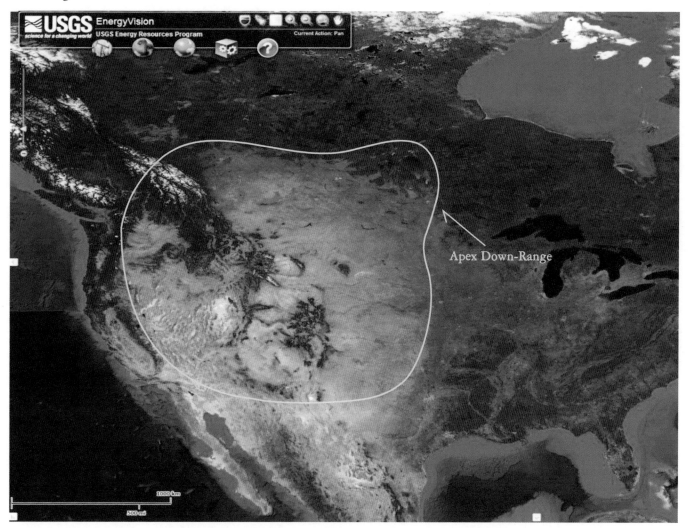

All of the ballistic evidence, is present in the crater to to show that Yellowstone super volcano is the result of a hypervelocity, very dense, large comet or asteroid that impacted in Nevada. It penetrated the earth at a oblique angle and plowed, subsurface, it's way northeast down and below Yellowstone.

Puzzle Piece #4 - The Smoking Gun

On April 3, 2009 I received my copyright certificate from the U.S. Copyright office. Just one month later, what I consider to be a major smoking gun surfaced. A discovery, adding another layer of substantial evidence, indicating a massive impact took place. Nature Geoscience published on 24 May 2009 that John D West, Mattew J. Fouch, Jeffery B. Roth, and Linda T. Elkins-Tanton, had discovered a near vertical downward flow associated with a Lithospheric drip beneath the Great Basin. In essence what they found was a 30 to 60 mile diameter bore in the Nevada great basin, pulling the earths crust downward into the mantel. The drip extends over 310 miles deep. This downward drip is hard to explain because it is located on the North American plate where it should not be. This sinking is not associated at a plate boundary where one would expect to find a subduction zone, which would cause the subsidence.

" This feature originally termed the 'Nevada Cylinder' [20] *, is approximately 100km /62 miles in diameter, extends near-vertical from ~ 75km /46 miles depth to at least 500km/ 310 miles (sic), and is bottom tilted to the northeast. Near 500km /310 miles(sic) depth, the cylinder merges with a separate zone of high-velocity material, making resolution of a distinct cylinder difficult below this depth. Resolution tests indicate that the Great Basin drip is well resolved, and the tilt is not an artifact of the tomography process..... We estimate that the volume of the cylinder is approximately 1 - 4 million km³ / 621,371 - 2,485,484 cubic miles (sic)"* [7]

Suddenly, here was a definite key piece of the puzzle that showed up right after my initial documentation. A cylindrical bore 30 to 60 miles in diameter, and over 310 miles deep. Pulling the ground or earths crust downward, right at the spot that I claimed the impact took place, and descending in a direct line with Yellowstone super-volcano. I contacted John D. West and verified with him that from the top of the cylinder to the bottom, the bottom was tilted and descending northeast towards Yellowstone. He replied, *"yes it was, but they could not make any ties between Yellowstone and the drip."* The reason being, is that the bottom of the drip is beyond the reach of their instruments where the cylinder merges with a separate zone of high-velocity material. The bottom that they can see with the methods that they used was down 310 miles, and was short of Yellowstone. (see image next page)

The center of the drip is located at approximately 117.8°W 39°N in Nevada. This Lithospheric drip, right at this spot, is not just a coincidence of random tectonics. This is the impact spot, and the bore hole path that the object took into the mantel of the earth. With all the data contained in this book, what better explanation do you have? Random plate tectonics? Subduction from off the coast reaching in pulling the mantel down? What makes more sense to you? All the data thus far, clearly defines a celestial impact crater without equal on the earth to date.

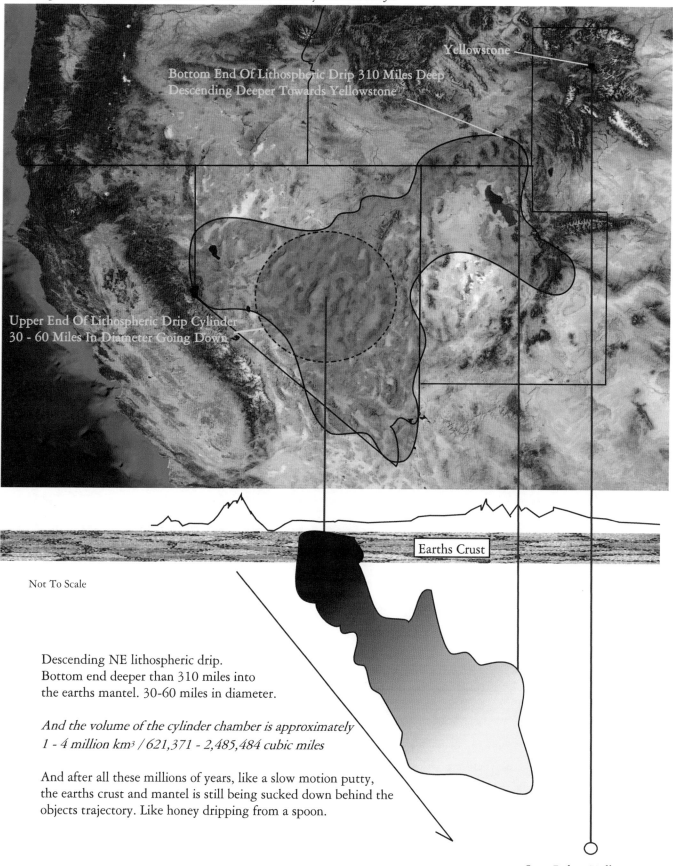

Base Image Credit - NASA The Blue Marble - Annotations Added by R.L. McGuire Jr.

Yellowstone

Bottom End Of Lithospheric Drip 310 Miles Deep
Descending Deeper Towards Yellowstone

Upper End Of Lithospheric Drip Cylinder
30 - 60 Miles In Diameter Going Down

Earths Crust

Not To Scale

Descending NE lithospheric drip.
Bottom end deeper than 310 miles into
the earths mantel. 30-60 miles in diameter.

*And the volume of the cylinder chamber is approximately
1 - 4 million km³ / 621,371 - 2,485,484 cubic miles*

And after all these millions of years, like a slow motion putty,
the earths crust and mantel is still being sucked down behind the
objects trajectory. Like honey dripping from a spoon.

Spot Below Yellowstone

Yellowstone's magma chamber extends further than previously thought......

"The huge column of molten rock that feeds Yellowstone's "supervolcano" dives deeper and fills a magma chamber 20 percent bigger than previous estimates, scientists say....

The finding, based on the most detailed model yet of the region's geologic plumbing, suggests that Yellowstone's magma chamber contains even more fuel for a future "supereruption" than anyone had suspected. The model shows that a 45-mile-wide (72-kilometer-wide) plume of hot, molten rock rises to feed the supervolcano from at least 410 miles (660 kilometers) beneath Earth's surface. "- Richard A. Lovett National Geographic News Published December 15, 2009

"Yellowstone is North America's largest volcanic field, produced by a "hotspot" -- a gigantic plume of hot and molten rock -- that begins at least 400 miles beneath Earth's surface and rises to 30 miles underground, where it widens to about 300 miles across." ScienceDaily (Nov. 8, 2007)

With the Lithospheric drip descending downward towards Yellowstone, with the measurable bottom at 310 miles deep, and Yellowstone further to the Northeast.... Yellowstone's magma chamber extending deeper, 410 miles is consistent with the objects decent path down and past Yellowstone.

"But a preliminary study by other researchers suggests Yellowstone's plume goes deeper than 410 miles, ballooning below that depth into a wider zone of hot rock that extends at least 620 miles deep.....

Smith says, "it wouldn't surprise me" if the plume extends even deeper, perhaps originating from the core-mantle boundary some 1,800 miles deep. -The studies were led by Robert B. Smith, research professor and professor emeritus of geophysics at the University of Utah and coordinating scientist for the Yellowstone Volcano Observatory." ScienceDaily (Dec. 14, 2009)

Again all of the data here in this book makes logical sense, that a celestial object created a conduit, as it bored its way deeper into the earth. What is needed now, is for the scientific community to actually produce a computer model of the impact. Using the data from the crater size, trajectory, humping, the lithospheric drip data, bore diameter, crater diameter, angle of attack of the object, densities, velocity, etc., and see if the model produces a working crater that matches the Yellowstone impact. That modeling, in and of itself will produce worlds of data in explaining the shear power it took to create a crater of this magnitude. It would also answer the main question that I have.... How big was the object that hit the earth? That is what I want to know, but without computer modeling, I can only guess it's size, velocity, and density. We have a bore hole 30 - 60 miles in diameter .... So what was the actual size of the object, taking into account the blast radius around object as it bored it's way down? If another object of this size was discovered heading towards earth, would we be able to stop or deflect it? Or would it just be time to say goodbye?

But not only that....

Does it match? Can the crater be verified through computer modeling?

Puzzle Piece #5 - When, K-T or Not?

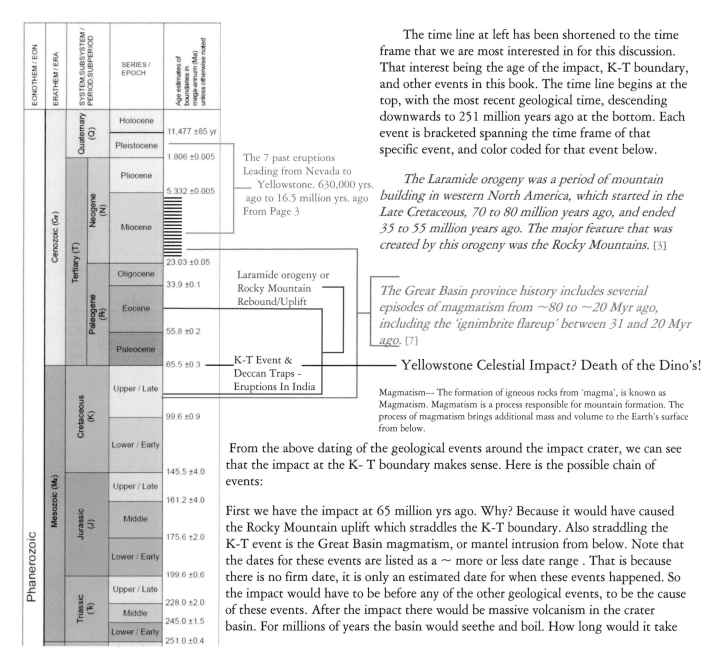

The 7 past eruptions Leading from Nevada to Yellowstone. 630,000 yrs. ago to 16.5 million yrs. ago From Page 3

Laramide orogeny or Rocky Mountain Rebound/Uplift

K-T Event & Deccan Traps - Eruptions In India

The time line at left has been shortened to the time frame that we are most interested in for this discussion. That interest being the age of the impact, K-T boundary, and other events in this book. The time line begins at the top, with the most recent geological time, descending downwards to 251 million years ago at the bottom. Each event is bracketed spanning the time frame of that specific event, and color coded for that event below.

The Laramide orogeny was a period of mountain building in western North America, which started in the Late Cretaceous, 70 to 80 million years ago, and ended 35 to 55 million years ago. The major feature that was created by this orogeny was the Rocky Mountains. [3]

The Great Basin province history includes several episodes of magmatism from ~80 to ~20 Myr ago, including the 'ignimbrite flareup' between 31 and 20 Myr ago. [7]

Yellowstone Celestial Impact? Death of the Dino's!

Magmatism— The formation of igneous rocks from 'magma', is known as Magmatism. Magmatism is a process responsible for mountain formation. The process of magmatism brings additional mass and volume to the Earth's surface from below.

From the above dating of the geological events around the impact crater, we can see that the impact at the K- T boundary makes sense. Here is the possible chain of events:

First we have the impact at 65 million yrs ago. Why? Because it would have caused the Rocky Mountain uplift which straddles the K-T boundary. Also straddling the K-T event is the Great Basin magmatism, or mantel intrusion from below. Note that the dates for these events are listed as a ~ more or less date range . That is because there is no firm date, it is only an estimated date for when these events happened. So the impact would have to be before any of the other geological events, to be the cause of these events. After the impact there would be massive volcanism in the crater basin. For millions of years the basin would seethe and boil. How long would it take for the impact basin to settle down from the up-welling of mantel and volcanism? How long to settle down enough to see the first of the trailing bubble eruptions reaching the surface? Well... according to the dated time line of events, the Great Basin province history includes several episodes of magmatism until ~20 Myr ago, so from the impact at the K-T boundary, 65 million years ago, to about 20 million years ago. Then the basin was calm enough, for about 4 million years, before the first trailing bubble eruption reached the surface and left it's mark on the surface. The first bubble eruption being the ~16.5 to 15 million year ago eruption. So looking at the time line again, it makes sense that the eruptions leading to the current Yellowstone site, happened after the basin settled down, after the Great Basin magmatism.

Note: In the time line, the Laramide orogeny, the Great Basin magmatism, and the K-T event would all have to start at the same time as the impact. The discrepancies in each of the estimated events beginning time frame, are just that... dating estimates and/or discrepancies.

So was this the actual K-T event? Is it just a coincidence that the Rocky Mountain uplift, and the Great Basin magmatism began right at the K-T boundary, and then lasted several million years after the event? Is it a coincidence that the Deccan Traps began at the same time? Is the impact the right size? It is definitely large enough to have been a major extinction event. For there is no larger impact site, currently on the earth, that has been identified. There is even today, the debate that the K-T event has not yet been identified. Some disagree that the, Chicxulub crater in Mexico was large enough. Others insist that the iridium layer from the Chicxulub has a different isotope signature from that found elsewhere in the K-T boundary. Was the Chicxulub crater a smaller fragment of the Yellowstone object?

Is this more evidence of a Shoemaker -Levy type event here on earth?.....

"The end-Cretaceous mass extinction has been attributed by most to a single asteroid impact at Chicxulub on the Yucatán Peninsula, Mexico. The discovery of a second smaller crater with a similar age at Boltysh in the Ukraine has raised the possibility that a shower of asteroids or comets impacted Earth close to the Cretaceous-Paleogene (K-Pg) boundary." [8]

An asteroid or comet big enough to produce the Yellowstone impact crater was more than likely accompanied by smaller fragments. Pelting the earth, like a blast from a shotgun. The main slug being the Yellowstone object, surrounded by smaller pellets.

"Asteroid Impact 65 Million Years Ago Triggered A Global Hail Of Carbon Beads......
The asteroid presumed to have wiped out the dinosaurs struck the Earth with such force that carbon deep in the Earth's crust liquefied, rocketed skyward, and formed tiny airborne beads that blanketed the planet, say scientists from the U.S., U.K., Italy, and New Zealand in this month's Geology. The beads, known to geologists as carbon cenospheres, cannot be formed through the combustion of plant matter, contradicting a hypothesis that the cenospheres are the charred remains of an Earth on fire." - Science Daily 5 May 2008

Where is the crater that produced the above carbon cenosphers that blanketed the earth? The Yellowstone impact was most assuredly of the magnitude needed to produce that effect. The K-T event, from what I can see, was not a one item event, but a culmination of events. From the choking gas, world wide volcanism, blocked sun light, loss of food, a fallout winter, all would have been the result from a impact the size of the Yellowstone celestial object.

So....... What now?

What Now

In summery we have all of the following evidence for a celestial impact more massive than any other on earth. A impact that even today has left a gapping wound still churning from the violence of that attack. 65 million years ago an attack on earth. One that created..., Yellowstone Super Volcano .... The Remnants of a Celestial Impact :

* The physical features of a complex impact crater.
* Visual satellite images showing expected geological, elevation, and physical features matching existing craters and patterns on the moon.
* Double ring crater rim 14,505 ft. above sea level with a lower outer ring.
* A impact basin dropping to -282 ft. below sea level, and an inner crater basin area of 1,436,190 miles^2
  a total crater area including the outer second rim of 1,602,045.8 miles^2 .
* A central rebound/uplift mountainous area. The forces to lift such a large area, geologists are at a loss to explain.
* An extensive geothermally active crater basin.
* The Great Basin magmatism spanning some 45 million yrs., with the intrusion of mantel material from below the earths crust.
* Magmatism or magmatic flows spreading north into Alaska and well south of Mexico. Even as far south as Guatemala.
* Multiple volcanic, 'bubble eruptions', in a direct line leading towards Yellowstone from the impact point beginning some 16.5 million yrs ago.
* The physical features of a hyper-velocity impact at an oblique angle with ground humping.
* A pear shape with the apex down range indicative of a high velocity impact.
* Evidence of shocked quartz in a zone extending out into the pacific ocean, nano-diamonds, iridium, glass spherules, and carbon cenospheres.... all evidence of a impact.
* A lithospheric cylindrical drip located right at impact, with a bore of 30 - 60 miles in diameter, extending over 310 miles deep, descending and tilted from top to bottom in a direct line towards Yellowstone. With a volume of 2,485,484 cubic miles.
* Yellowstone itself having a magma chamber that extends deeper than the lithospheric drip + 410 miles and a volume of magma which is hard to imagine. The magma chamber depth having the proper relationship consistent with a descending projectile as it relates to the bottom depth of the lithospheric drip.
* A geological time line matching the events as they progressed at the K-T boundary. And in the proper sequential order...
  1. Impact & K-T Boundary
  2. Magmatism & Rebound
  3. Settling of the Basin
  4. Bubble Eruptions inline with trajectory and proper time relation, with the closest bubble to the surface being the first to erupt, and the deepest bubble being the last to rise to the surface to erupt.

Another potential crater site that is worth looking at, would be a satellite image of the north end of South America. In comparison to the Yellowstone crater, which has the distinct features of a high velocity oblique angle impact. The South American image shows a crater with a double ring rim to the west. A distinct crater basin area. A nearly perfect symmetrical ring, and appears to be a low velocity coupling at a vertical attack angle. The whole basin drainage is from the crater ring inwards, tilting to the east where it drains into the Atlantic ocean.

The crater inner basin covers approximately 3,307 km/1,900 miles in diameter. And covers an area of 7,849,382 km^2 or 3,030,663 miles^2 (red area overlay). In both case random plate tectonics cannot explain nearly symmetrical ringed basins, with the features of impact craters.   Image Credit - NASA - The Blue Marble

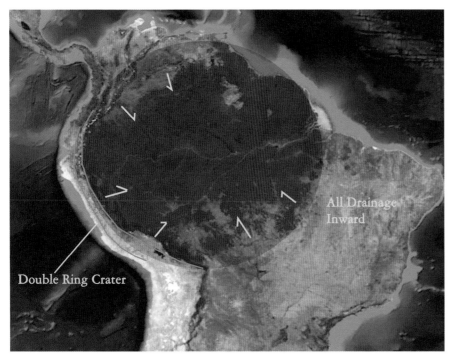

Double Ring Crater

All Drainage Inward

Is this further evidence of a shower of objects that hit the earth? Is it the same age as the Chicxulub, Yellowstone, and other impacts? Is this one even a impact crater.... If so I am naming it the, "Ricky II South America Impact" site. Yes, I have a sense of humor.

From all the data here, it is clear that Yellowstone, and a major section of the United States, is a massive impact crater. Further ground data, and input from the scientific community at large will either confirm or dismiss this. One has to look at all that was discussed here, and realize that this answers some major questions that plate tectonics just cannot explain. IE: The massive uplift of the Colorado Plateau provinces and a logical relationship between the Great Basin Lithospheric drip and Yellowstone.

Giving a better explanation for the existence of the Lithospheric drip's location in the middle of a plate where none should exist. A better explanation for the existence of Yellowstone's location, and the mechanics of the volcano, and many other puzzle pieces in answering the formation and geology of the western United States.

It also raises a lot more questions.... The Chicxulub crater on the Yucatan peninsula is thought to be the event that ended the Cretaceous period, and was the death of dinosaurs, while others contend it was not. So which one was the KT event? Would one expect to find the KT boundary in the actual blast basin of the western U.S.? Or only outside the basin? Leaving so many other questions. Is the north end of South American a large impact crater also? We have evidence of craters on the moon this massive...... Why not here on earth also?

Discovery is still alive, and will be forever. For no man has seen it all. Under every rock, around every bend there lies in wait something new to discover. That is what has made this adventure so enjoyable. There is enough overwhelming physical evidence here that this is a celestial impact, that I once again claim the right of discovery to name the impact site, the: "Richard L. McGuire Jr. Yellowstone Impact".

Though there will be many that will contend why this is not possible, with counter theory after theory, you have been given the facts as clearly as I could give them. And it is for you to look, and then wonder at the marvel that is in your own backyard. For when you look out, where ever you live in the western United States, you can say, ... I am living in a impact crater! And the next time you visit Yellowstone, you can look with new awe at the true wonder that it is. It is no wonder this impact crater has not been seen until now. For it truly is such a massive crater, that from our puny view point walking the earth, one can only look out and ask... What crater?

I leave you with this one parting photo, and say to the scientific community... it is now up to you, who are more competent than I, to go forth and find out the truth of what I claim. And then write the new story of our geological past!

Let The Debate Begin!

Parting Photo .... The Anaxagoras moon crater from page 9, laid over the Yellowstone Impact as a semi-transparent layer. The only thing I did was make it a transparent layer, and scale the image. As you can see, the physical features of the two images are a perfect match!

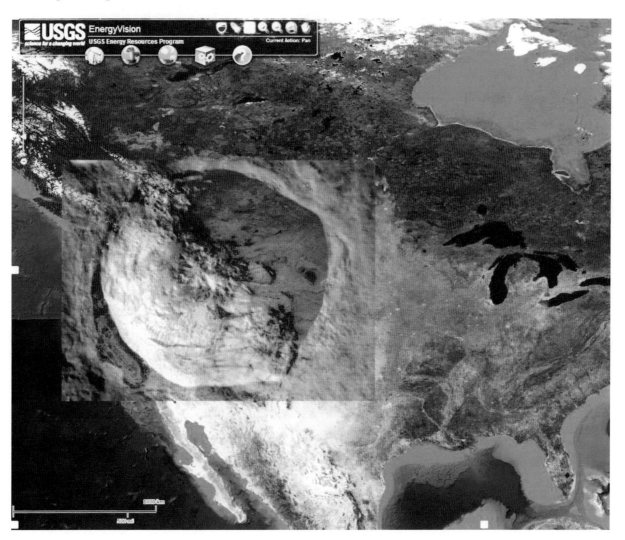

References

Page 3   1   Description : NASA Godard LOLA image centers on the South Pole-Aitken (SPA) basin

Page 6   2   NASA Landsat 7 Education Activity Finding Impact Craters with Landsat

Page 10 & 23   3   USGS CVO Menu - America's Volcanic Past - Colorado
Excerpt from: U.S. National Park Service, Black Canyon of the Gunnison National Park Website, 2002

Page 12 & 17   4   Tracking the Course of Evolution THE K-T EXTINCTION by Richard Cowen
Richard Cowen Ph.D., Cambridge (1966) Senior Lecturer Emeritus;
University of California Academic Senate Distinguished Teaching Award Recipient, 1993

Page 14   5   Journal of Volcanology and Geothermal Research 131 (2004) 397-410
Evidence for gas and magmatic sources beneath the Yellowstone volcanic field from seismic tomographic imaging
Stephan Husen, Robert B. Smith, Gregory P. Waite
Department of Geology and Geophysics, University of Utah, Salt Lake City, UT, USA
Received 10 June 2003; accepted 20 November 2003

Page 19   6   Lunar and Planetary Science XXVIII 1787.PDF -
Forming The South-Pole Aitkens Basin: The Extream Games P.H. Schultz.
Dept. - Geological Sciences Brown University, Providence, RI 02912

Page 20 & 23   7   Nature Geoscience published online : 24 May 2009 DOI: 10.1038/NGEO0526
Vertical mantle flow associated with a lithosphric drip beneath the Great Basin
John D. West, Mattew J. Fouch, Jeffery B. Roth, and Linda T. Elkins-Tanton

Page 20   20   *Nevada Cylinder - Roth, J.B., Fouch,M.J., James,D.E., & Carllson,R.W.*
Three-dimensional seismic velocity structure of the northwestern United States,
Geophys. Res. Lett. 35,L15304 (2008)

Page 24   8   Geology geology.gsapubs.org doi: 10.1130/G31034.1 v. 38 no. 9 p. 835-838
Two large meteorite impacts at the Cretaceous-Paleogene boundary
David Jolley1,*, Iain Gilmour2, Eugene Gurov3, Simon Kelley2 and Jonathan Watson2

Richard L. McGuire Jr.

I grew up in Santa Barbara California, loving the outdoors.
I was always up for a hike, exploring just to see what was
around the next bend. Sometimes it was hard to call it quits,
and I would often find myself trekking home way after dark.

When I was older, I was traipsing around the Sierra Mountains.
Convict lake just outside of Mammoth Lakes was a favorite spot.
Hiking up from Convict Lake to Lake Genevieve, and from there
to Edith Lake. Camping and fly fishing for trout made for some
wonderful Memories.

The drive up the back side of the Sierra Mountains on
U.S. Route 395 is just breath taking. Seeing Mt. Whitney rising
up on your left as you head north from Mojave California, is
something everyone should experience once in their life.

These are the things growing up in life that made me want to look around the next bend. To discover what was under,
around, and over the next hill. In a sense, one could say I was a frustrated geologist, and I should have pursued that
occupation. Writing this book has been an adventure itself. Finding each new piece of the puzzle was not only a
challenge, but it was very satisfying when the piece fit.

My hope is that you will discover for yourself, the joy of finding answers to things that have not yet been discovered.
For no man has seen it all, and the world is full of wonderful things just waiting for you to bring them out of the
darkness and into the light.

I currently live in the country just outside of Aberdeen Washington. A twenty minute drive puts me in the Olympic
Rain Forest where I enjoy hiking and photography. To see some of my photography visit RicksCanvas.com

Robert

Its Good to have you
as a friend !

Richard L. McGuire Jr.

Made in the USA
Charleston, SC
01 July 2012